# Silkie Chickens: The Complete Owner's Guide

The Must Have Guide for Anyone Passionate about Owning, Breeding or Showing Silkies

By: Ruth L. Corbin

Copyright © 2020 by Ruth L. Corbin

All rights reserved. No part of this publication may be reproduced, distributed, or transmitted in any form or by any means, including photocopying, recording, or other electronic or mechanical methods, without the prior written permission of the publisher, except in the case of brief quotations embodied in critical reviews and certain other noncommercial uses permitted by copyright law. For permission requests, write to the publisher at the address below:

Heath Publishing Company
1525 Carlos Dr.
Greenville, NC 27834

# Table of Contents

Introduction ............................................................................. 1

Chapter 1: History of Silkie Chickens ............................................... 3

Chapter 2: Characteristics and Traits of Silkie Chickens ................. 8

Chapter 3: Housing and Feeding Silkie Chickens ......................... 13

Chapter 4: Silkie Chickens as Pets and Show Birds ..................... 21

Chapter 5: Breeding Silkie Chickens ............................................. 38

Chapter 6: Silkie Healthcare ......................................................... 46

Chapter 7: Silkie Chickens for Eggs and Meat .............................. 53

Chapter 8: Conclusion – Are Silkies the Breed for You? ............... 59

# Introduction

Whether you already own Silkie chickens or you are considering owning them, this is the book for you. This is not a book on general chicken care, which is well covered by other sources. After reading this book you will be an expert on Silkie chickens. You will be able to:

- Describe the history of the Silkie breed in stunning detail
- Know what characteristics and traits to expect from your Silkies
- Optimize housing and feeding for your Silkies
- Keep your Silkies safe within a mixed breed flock
- Selectively breed Silkie chickens to produce your own prize-winning show birds
- Safely keep Silkie chickens as family pets
- Use your Silkie chickens for fresh eggs and even meat
- Hatch Silkie eggs both naturally and through incubation
- Determine whether your local climate is suitable for Silkie chickens
- Easily prevent a deadly disease which Silkies are highly susceptible to
- Healthcare for Silkie chickens
- Silkie chickens for medicinal uses
- Weigh the pros and cons of the Silkie breed
- Make an educated decision on whether Silkies are the breed for you

- Provide customized, expert care for your Silkie chickens
- Win shows and exhibitions with your Silkie show birds

I know you are excited to get started, so let's jump right into it!

# Chapter 1: History of Silkie Chickens

Silkie chickens have a long and proud history, although their exact origins remain unknown. They are thought to have originated in China or elsewhere in southeast Asia. One theory is that the Silkie chicken originated during the Han Dynasty in China (206 BC – 220 AD). Because of the numerous special and unique genes exhibited by the Silkie, the creation of the breed would have required a great number of breeders with a great number of birds over many human generations. This theory proposes that the breed was created by the wealthy and selectively bred to enhance its regal appearance. Other theories suggest that the breed may have been originally developed in Japan or India before making its way to China. The location of origin is up for debate, but the creation by selective breeding is not disputed. Experts agree that highly orchestrated selective breeding techniques must have been used to create the unique and beautiful Silkie breed.

There is a great deal of information about the Silkie in ancient Chinese writings. Ancient cultures believed the Silkie chicken to hold medicinal powers, and some still practice the belief to this day. They may look like some new fad breed of chicken, but Silkies are actually one of the few ancient heirloom chicken breeds that we still have around today. The Silkie chicken is one of the earliest examples of breeding exclusively for appearance in livestock. Many of the genes of the Silkie breed would have had to been dutifully and carefully bred to maintain. For example, the gene which gives the Silkie its unique internal and skin color involves the duplication and inversion of two choromonal segments. Other genes that could have only occurred by human intervention include the Silkie's feathers, comb, crest, muffs and beard, and the polydactyl gene (having five toes).

The first recorded western sighting of the Silkie chicken was by Marco Polo in 1298. He marveled over a chicken with black skin and "hair like a cat" on his travels to China. (Although it is unknown whether he saw the birds or just heard about them.) As trade routes between East and West were established, the Silkie chicken made its way to Europe. This is possibly where the Silkie received its name, as they were brought to Europe via the Silk Route. Another theory is that the city where Marco Polo first saw or heard of these birds, Kue-Lin-Fu, was wealthy from silk production. And yet another theory is that the name was given simply because their feathers have an extremely silky texture due to the lack of barbules.

Some Europeans didn't quite know what to think of the Silkies' extraordinary looks. Records show that in the Netherlands, they were sold as the product of crossing a rabbit and a chicken! Nevertheless, the Silkie inevitably became a popular breed of poultry across Europe. In 1598, Ulisse Aldrovandi, a writer and naturalist at the University of Bologna, Italy, published a comprehensive treatise on chickens which is still read and admired today. In it he discussed the "wool-bearing" chickens and described them as "clothed with hair like that of a black cat". Silkies were also written about by French historian Jean-Baptiste Du Halde in the 1700s, who specialized in Chinese history.

Silkie chickens were officially recognized by the Great Britain Poultry Club and the American Poultry Association in their first published standards. Silkie chickens were recognized as both large fowl and bantam varieties with listed Standards of Perfection for each by the Great Britain Poultry Club in 1865. Silkie chickens were recognized only as a bantam breed with listed Standards of Perfection by the American Poultry Association in 1874. It is interesting to note that each group recognizes a different "ideal" breed size.

By the early 1900s, Silkies were being exploited in travelling circuses and side-shows across the United States as 'freaks', described as "Chickens with fur instead of feathers." They were literally touted as a hybrid bird-mammal.

Actual circus sideshow banner. Brushed by artist "Butch" in 1896.

Silkies are now present in North America, Europe, Asia, and Australia. Their popularity, population and diversity spread globally. Unlike many of the heirloom chicken breeds, Silkie chickens managed to retain their popularity during the industrialization of the 20$^{th}$ Century. During this time the poultry business became dominated by a handful of large corporations and backyard flocks were no longer the norm. While most heirloom breeds were listed as endangered by groups such as the Rare Breeds Survival Trust in England and the American Minor Breeds Conservancy (now the Livestock Conservancy) in the United States, the Silkie chicken managed to stay off these watchlists.

Perhaps it is the unique characteristics and appeal of the breed that has withstood its popularity across many generations.

Silkies are now ranked in North America's top fifteen chicken breeds and are most commonly used in backyard flocks by hobbyist chicken owners. They are also commonly used as show birds due to their royal appearance. While most owners eat the eggs produced by their Silkies, this is not generally the primary purpose of ownership, as there are other breeds which would be much more productive in this area. While not a common practice in western culture, Silkies can be used for meat. The "black skinned chicken" is a delicacy in many Asian cultures.

Most breeders focus their programs on developing and enhancing the unique characteristics of the Silkie chicken, as it is the breed's handsomeness that generally attracts owners. The Silkie chickens seen today have changed somewhat in the last thirty to forty years. Their crests are larger and the feathering down the legs is more abundant than in their ancient counterparts. There are also a larger variety of colors available today. The original Silkies had white feathers only, but through meticulous breeding by some dedicated fanciers, additional colors are now available. However, modern day Silkies still closely resemble the ancient Silkie chicken breed which was created over 2,000 years ago.

# Chapter 2: Characteristics and Traits of Silkie Chickens

Silkie chickens are docile, and somewhat lazy in habits compared to other breeds. It is said that the Silkie is the only breed that could enjoy living inside an apartment, although I would recommend a rooftop coop! Silkies generally exhibit a friendly nature and make good family pets. They are relatively easy to manage, making Silkies a great option for inexperienced chicken owners. Even the roosters are known to be calm, friendly and docile. Silkies have a reputation for being one of the most tolerant breeds for young children and most enjoy sitting in your lap. They are very much the cuddly pet of the chicken world.

Silkies are unique looking chickens, which combined with their mild temperament is what makes them popular as pets and show birds. Silkies are regal in appearance and entirely different from all other breeds of fowl in that their feathers lack barbicels (the hooks that hold the feathers together). This makes their feathers look like fine hair or silk and gives the Silkie its characteristic fluffy appearance. Because Silkie chicken feathers are not held together with barbicels, they cannot fly. Another unique quality of Silkie chickens is that their heads are crested, looking somewhat like a 'pom-pom'.

The original Silkies were white, but they are now available in other colors including black, blue, grey and buff. Silkie chickens have black skin, faces and bones. Their flesh is a very dark grey-blue. Their combs (if present) are black or dark mulberry in color and walnut shaped. They have oval shaped turquoise earlobes and dark colored wattles. A red comb or wattle indicates that the chicken is not a pure breed Silkie. Silkie chickens' beaks are grey/blue in color, short and quite broad at the base. Silkies have large wondering black eyes which (maybe it's just me…) seem to always be following you when you are in their sight radius.

Silkie chickens' bodies are broad and stout with short backs and full breasts. They have five toes instead of the usual four toes found in chickens. At minimum the outer two toes should be feathered (unless their feet have been allowed to be wet for long periods of time). The feathered legs are short and wide set, with grey skin color. Silkie chickens can be standard or bearded varieties, with the bearded variety including a 'beard' and muffs.

While Silkie chickens are used for eggs, this is usually a secondary purpose for the hobbyist owner as they are somewhat poor layers numerically compared with other breeds. However, Silkie hens are excellent at sitting on eggs. Because of that characteristic broodiness, Silkies are often used as sitters to hatch the eggs of other breeds.

It is not common in western cultures to use Silkie chickens for meat, but if you can get past the black skin and dark meat, you certainly can eat them. Silkie chickens are eaten in many Asian cultures and considered a delicacy. Some cultures also believe that the "black-skinned chicken" holds medicinal powers.

Silkies are relatively small chickens. Interestingly, all Silkies in the United States and Canada are considered to be bantam, regardless of size. In the United Kingdom, large fowl Silkies should weigh around 4 pounds for the roosters and 3 pounds for the hens when mature. The bantams should weigh around 21 ounces (1.4 pounds) for roosters and 18 ounces (1.2 pounds) for hens when mature. In the United States the bantam Silkie should weigh around 36 ounces (2.3 pounds) for roosters and around 2 pounds for then hens.

Their small size is one of the reasons that Silkies are not well adapted to snowy weather, although they are more cold hardy than many other small or bantam breeds. If you live in a northern climate with frigid winters, the Silkie might not be the best breed for you, unless you are willing to invest in a heated chicken coop and take extra precautions to keep your Silkies safe and comfortable.

Silkie chickens also dislike being wet, because their feathers become soaked and do not repel water like other poultry breeds (due to the lack of barbicels). A drenched Silkie is truly a pathetic sight. You can dry them with a towel or even carefully blow dry them when they get wet. If you live in a climate with very high levels of rainfall, such as the tropics or the pacific northwest, Silkies may not be the best breed for you, unless you are willing to invest in a substantial coop or barn to keep them dry, including their feet. Silkie chickens tolerate heat reasonably well, making the breed an excellent choice in hot dry climates.

Silkies are a relatively low energy breed and are considered lazy by chicken standards. Because of this they don't require a great deal of space, making Silkies an ideal choice in cities and other places where space is at a premium. Just be sure to check your local ordinances if you live in a city or town limits. Rules are determined locally and differ from city to city. Many cities are allowing people to have a certain number of laying hens, while some are not. Very few cities allow roosters. Your best bet would be to speak with someone in the Planning or Zoning Department at your City Hall. If you live in the county (or other equivalent outside the city limits), this should not be an issue for you. (This references United States zoning laws.)

If you have more space and are able to give your Silkie chickens some room to roam, they will definitely do some foraging and supplement their feed. Just don't expect as much effort from them as you would with most other breeds allowed to free range. A benefit of being lazy is that Silkies don't tend to roam too far from home.

Silkie chickens are arguably the most interesting looking birds you can have in a backyard flock. Most people who have owned Silkies fall in love with the breed and continue to include them as part of their backyard flocks for years to come.

# Chapter 3: Housing and Feeding Silkie Chickens

As with any breed, Silkie chickens will need a coop with nesting boxes and a fenced area if you want to let them out of the coop and keep them contained. Some people say you should not free range Silkie chickens because they can get lost more easily or be taken by hawks. I haven't had a problem with Silkie chickens getting lost. I'm sure people have, but I think it is a myth that they get lost more than any other breed of chicken. Actually, I have found that a benefit of their laziness is that Silkies don't tend to wander too far from home.

The other fear, that Silkies can more easily be taken by hawks and other large birds, is true. However, the same is also true with any small or bantam chicken breed. In my opinion, some freedom to free range is beneficial for any chicken if possible, but this is a personal decision. Some people might feel like the risk outweighs the reward with smaller breeds. Other options include covered runs attached to the chicken coop and portable chicken tractors.

Silkie chickens are a low energy breed and are very docile compared to other breeds of chickens, even more so than most known "friendly breeds". If your Silkies are housed with more aggressive breeds they may be attacked by the others, who will tend to take their kindness as a weakness. If you do plan to have a mixed flock, observe the behavior of the chickens to make sure your Silkies are not being bullied. If you notice any serious issues, you may need to consider a separate coop for your Silkie buddies. They should do fine if house with other breeds known for low aggression, such as Buff Orpingtons, Cochins, and Plymouth Rocks.

Because they are so easy going, Silkies are less likely to try to escape than many other breeds. Silkie chickens cannot fly, so you don't have to worry about them flying over a fence. A three-foot high fence is tall enough to contain Silkies. Because they do not have any barbicels to hold their feathers together, their wings will not hold air. They can sometimes partially jump and fly for a few feet with the aid of what wing feathers they do have, but they cannot fly for extended lengths like some other breeds.

Because Silkies are a small, low energy breed, they don't require a ton of space. A five-foot by seven-foot coop is adequate for six chickens, or up to twelve if there is a large run. However, be sure not to overcrowd. Smaller sized nesting boxes in the coop are appropriate. The perches and nesting boxes in the coop should not be more than six to eight inches off the floor, since Silkies can't fly.

A small run or chicken tractor is plenty adequate for Silkie chickens to get out and move around. Although you can get away without one. Silkies are one of the most contented breeds in confinement. If you have an outside area for your Silkies, putting down sand in that area will help keep the Silkie's feet dry, protecting their foot feathering. Since Silkies really are sweet and cuddly pets; a lot of owners even bring them into the house to hang out with the family some! Silkies are the only true couch potatoes of the chicken world, making them a favorite in families with children.

Silkies do not tolerate extreme cold temperatures. They are not as sensitive as some small and bantam breeds, but Silkie chickens are not cold hardy by any means. There is some misinformation on the internet around this topic. I think the confusion is in comparing Silkies with the most cold sensitive chicken breeds in existence. No, they are not that bad, but Silkie chickens are not cold hardy like say the Buff Orpington, Australorp, or Rhode Island Red. These breeds are all good choices in places with frigid winters, tolerating below zero temperatures. Dig a little deeper to find stories about people disappointed that their Silkie chickens died over the winter and they have never had that problem keeping other breeds of chicken. Those stories speak for themselves. Silkie chickens can tolerate some cold, but they are NOT cold hardy for extreme winter environments. Bearded varieties are just slightly more cold hardy than standard Silkies.

If you really want Silkie chickens and you live in a climate with frigid winters, you could invest in a heated barn. That's a lot for chickens but you may be dedicated and have the money! Make sure the barn is ventilated and just heat it to above freezing, no need to make it too warm. If you go this route, make sure your heating system is safe and up to code. Don't use unattended heat lamps or heating implements of a similar nature to keep your chickens warm. The chances of a fire are too real and just not worth the risk.

If you live in a tropical climate or somewhere with high levels of rainfall, investing in a well-ventilated barn with a real floor would be ideal if you are intent on keeping Silkie chickens. At minimum their chicken coop should be under a shelter of some type. As mentioned in the previous chapter, Silkies HATE being wet, including their feet. Their feathers do not allow the rain to run off as with other breeds of fowl, due to the lack of barbicels in the feathers. Their feathers get wet more like your hair does, and the Silkie chicken left in the rain will be a dripping wet, miserable mess. It is especially important to keep young Silkie chicks dry. Wet crests in young Silkies are a prolific source of colds.

Silkie chickens are very tolerant of the heat. This makes the breed a great choice in hot, dry climates, in addition to temperate climates.

Silkies will not need any special forms of housing in these environments, which they are naturally adapted to.

Because Silkie chickens are a small breed and lazy in habits, they must not be overfed. Feed sparingly and keep them alert and active. An overfed Silkie will likely become fat, stop moving around, and the hens may begin to lay eggs without shells.

As with any small or bantam breed, fresh greens and protein should make up a higher percentage of their diets as compared with standard breeds. Their need for the additional protein not provided by commercial feed can be met by allowing Silkies to free-range for worms and bugs during spring and summer months. During the winter (or anytime they are not allowed to free-range) Silkies can be supplemented with soy, milk, fish, etc. to get the extra protein they need. For greens you can simply let them out in an area filled with grass and weeds to peck at. Or you can give them scraps of green vegetables from the garden and the kitchen.

Besides the need to supplement commercial feed, feeding Silkies is like feeding any other chickens. Silkie chicks should receive a starter chick feed (20 – 22 percent protein) until about six weeks of age. Then they should be switched to a pullet grower feed (14 – 16 percent protein) until about 20 weeks of age. Then you can start feeding them layer feed (15 – 18 percent protein). Choose layer crumbles instead of layer pellets. Since Silkies are small the layer crumbles are easier for them to eat. You can mix cracked corn in with the layer feed if you like. This is a good idea during winter months because it helps the chickens stay warm.

Most people buy pre-mixed commercial chicken feed. You can get this at any agricultural supply store. You can mix your own chicken feed if you desire and have the time. You might want to if you are raising Silkies as show birds. As with people they are healthier and more beautiful if they have a good diet. A very strict time-tested method is feeding chick starter feed with hard-boiled egg chopped fine for the first three weeks, then adding rolled oats to the mixture until they are eight weeks old. After they are eight weeks old, they are fed a mash made of coarse cornmeal and rolled oatmeal soaked overnight in skim milk in the morning. Midday they are given boiled potatoes, carrots and beets, and for dinner they are given whole wheat. (Never give chickens raw potatoes). Wheat bran can be available at all times, but never give more of the other rations than they can eat up clean. Do not feed corn to show birds except on cold winter nights. If the birds are able to have free range, natural bugs and forage will also be beneficial.

Adding any of these natural food sources will be good for your Silkies and they will appreciate it too. It does not have to be all store-bought feed or all fresh food, it can be a combination. Most people provide commercial feed as the staple, and supplement with fresh food. There is also a difference in quality between brands of commercial feed. Be sure to check the protein content and compare with the guidelines above. Read the ingredients so you know what you are feeding your chickens.

Even a small enclosure outside the chicken coop is sufficient for your Silkies to find worms and bugs to eat. Silkies also do well as city birds so no problem there. If you live in the city where they will remain in the chicken coop, you can still supplement by purchasing worms for your chickens and sharing scraps of fresh veggies. Worms can be purchased at any tackle shop and many pet stores sell mealworms. Providing grit for your Silkies is also needed if they are not allowed to free-range. If allowed to free-range they will pick up natural pieces of grit.

Silkie chickens can benefit from iodine supplements when they are going through a moulting period. (During moulting chickens lose feathers and regrow them. They usually stop laying eggs during this time to build their energy reserves.) The iodine supplement will help the Silkies grow back an abundance of beautiful plumage. You can give them a seaweed supplement that contains iodine. Or you can put two teaspoons of tincture iodine per gallon of water (or use potassium iodide tablets dissolved in water) so the chickens will get the iodine when they drink. Be sure to measure and not put too much iodine in the water, which could kill your chickens.

If you aren't set-up for chickens yet, you can build a chicken coop yourself and save a lot of money if you have basic carpentry skills. However, working out dimensions, materials, insulation, ventilation, lighting, positioning, nesting, perches, waste collection and protection from the elements and predators can seem complicated. This can lead you to purchasing an expensive pre-made chicken coop just to save yourself the headache of figuring it all out. Fortunately, you can get an easy to follow guide to building your own chicken coop. You get multiple blueprints in the guide so you can customize the coop to your size and budget needs. You also get supply lists and detailed instructions, so you know exactly what to buy and what steps to take. This guide has helped thousands of people build their own affordable backyard chicken coops. You can do the same by getting your guide at https://www.buildingachickencoop.com/?hop=chicken121 (this is an affiliate link).

## Chapter 4: Silkie Chickens as Pets and Show Birds

Silkie chickens have a calm, friendly disposition. They are well known as one of the best breeds to keep as pets (and by many as the absolute best). It helps that they are fluffy and enjoy sitting in your lap! If you have children and want to have backyard chickens, Silkies are a great choice. As compared with other breeds, Silkie chickens are very friendly and have laid back personalities.

In addition to their friendly personalities, many people like to keep Silkies as pets because they are beautiful and unique looking birds. Their regal appearance makes them look like royalty of the chicken world! Some people enjoy showing their Silkie chickens at exhibitions, shows and fairs, and even win cash and prizes for doing so.

If you are going to show Silkies, you need high quality birds. The American Poultry Club recognizes Silkie chickens only as bantams while the British Poultry Club recognizes both bantam and standard varieties. Sometimes you will see Silkies advertised as small, medium, or large varieties. The "large" refers to the standard size Silkies according to British Standards of Perfection, while the "small" refers to the bantam size Silkie according to British Standards of Perfection. The "medium" size refers to the bantam size recognized by the American Standards of Perfection (which is the only size recognized by the American standard). All Silkies are fairly small, with full grown Silkie chickens ranging from approximately one and a half pounds to four pounds for roosters and one pound to three pounds for hens.

A show quality Silkie absolutely must have five toes, silky feathers (lacking barbicels) over most of the body, feathered legs, and dark blue to black skin. According to both American and British Standards of Perfection, any Silkie without these qualities must be disqualified. Red combs, faces, or wattles will also get a Silkie disqualified according to American Standards, while this is a serious defect that will cause point deduction according to British Standards, but not disqualification.

Having a single type comb is cause for disqualification by British Standards and is a serious defect by American Standards. (The single comb is what people typically think of when envisioning a comb, being a thin tall comb with five or six points that begin at the base of the beak and ends at the back of the head). Walnut type combs are ideal according to American Standards, while walnut, strawberry, and cushion types are all equally acceptable according to British Standards. All of these are the flatter type combs which sit close to the face.

The toes of Silkie chickens should be feathered. Lack of feathering on the toes is a defect according to both standards. Therefore, you should keep Silkie's feet dry (although some are just born without the gene for feathered feet).

A show quality Silkie should exhibit the correct texture. We want the texture of the fluffy substitute for feathers to be long, and as silky as possible, soft in texture and with as little hard quill as possible. Even the long wing feathers should be devoid of hard feathering. The center and outer toes should be feathered, as well as the shanks, but without any trace of vulture hock. Vulture hock is hard feathering in the hock which points down and is a disqualification for Silkies according to both standards.

In addition to their fluffy feathers, the crest helps give Silkies their unique and regal appearance. A poor crest is a defect according to both sets of standards. I like the way the British Standard describes the Silkie crest, "A pompom like a powder puff, which in the male will have shiny streamer feathers pointing backwards." Silkie colors recognized by the American Standard include white, black, blue, partridge, buff, gray, splash, self-blue (lavender), and paint. Colors recognized by the British Standard include white, black, blue, gold, and partridge.

White Silkies should be a brilliant white color in all sections. Black Silkies should have a lustrous greenish black color, with the under color being dull black in dark-legged varieties. The black should be even although a small amount of color is permissible in the hackle, but not desirable. Blue Silkies should have an even shade of clear bluish plumage and a glossy black head. The neck, back, tail, and breast feathers should have a lacing of black. The under color of all sections should be a bluish slate. The blue should be an even color with no patchiness or splashing.

Partridge Silkie males should have a head that is a lustrous, rich red. The neck should be a lustrous greenish black with a narrow lacing of medium shade, rich, brilliant red, with a black shaft. The front of the neck should be black. The back should be lustrous greenish black with a narrow facing of a medium shade of rich brilliant red. Rich brilliant red should predominate the surface of the upper back, the saddle matching with the hackle in color. The main tail should be black with main and lesser sickles being greenish black and coverts being greenish black laced with medium shades of rich brilliant red. The wings should be black in the front, with the bows a medium shade of rich brilliant red, and coverts a lustrous greenish black, forming a distinct wing bar of this color across the entire wing when folded. The breast should be a lustrous, greenish black. The body should be covered in black fluff slightly tinged with red. In the female, the patterns are the same, but the red coloring is a deep reddish bay, rather than a lustrous or rich brilliant red. British Standards refer to the red coloring as dark orange.

Buff Silkies should have an even shade of rich golden buff throughout the surface. The British Standard recognizes gold varieties rather than buff. Gold Silkies should have a bright, even shade of gold throughout, with darker feathers permissible in the tail.

Gray Silkie males should have dark gray heads and gray Silkie females should have chinchilla gray heads. The hackle in both sexes should be light gray streaked with darker gray. The back should be an even shade of chinchilla gray, and the saddle in the male should match the hackle in color. The tail should be an even shade of chinchilla gray. The wings should be an even shade of chinchilla gray for the shoulders, fronts, bows and coverts, while the primaries should be a slatey gray with a center shafting of a darker shade and the secondaries should be a slatey gray. The breasts should be light gray. The body and stern should be an even shade of chinchilla gray. The shanks and toes should be a slatey blue. The under color should be a smoky gray of a shade not darker than the top color.

Self-blue Silkies, also known as lavender Silkies, should be an even shade of light blue or lavender all over. Paint Silkies should be white with black spots or polka dots. They are basically marked like Dalmatians. Splash Silkies have white or pale blue plumage with irregular patches of black or dark blue color. Rather than solid polka dots like with paint varieties, the markings look like light splashes throughout. Splash Silkies are not recognized by British Standards and they are the most recent addition to American Standards, having been added in the year 2000.

The Silkie body should be of moderate length, broad, deep and well-rounded from breastbone to stern and let down well between the legs. In the female the under fluff should nearly touch the ground. The breast should be carried forward, very full, well rounded and of great depth and width. The back should be short, broad, and quite rounded in its entire length, rising gradually from the middle of the back toward the tail. The Silkie's neck should be short and gracefully arched, with a very full hackle flowing well over the shoulders. The head should be moderately small, short, and should be carried so that a line drawn parallel with the tip of the tail would bisect the comb.

The Silkie's tail should be short, very shredded at the ends, well spread at the base, and filled underneath with an abundance of soft feathers. The wings should be medium size, closely folded, and nearly horizontal, well above the lower thighs ending short of the stern. The wings should be ragged with some of the flight feathers hanging down, almost looking tattered. Split wings (where there is a gap between the primaries and secondaries) are a defect.

Silkie legs should be short and stout, set well apart, and should appear straight when viewed from the front. Lower thighs should be short, stout at top, tapering to hocks, and abundantly feathered. Hocks should be covered with soft and silky feathers curving inwards. Shanks should be rather short, stout in bone, well feathered on the outer sides with silky plumage, the upper part growing out from thigh plumage and continuing into foot feathering. Males only have spurs medium size in length, just above the fifth toe. There must be five toes. The three front toes should be straight, well and evenly spread, while the hind toe should be double, the normal toe in natural position and the extra toe placed above. British Standards specify that each toe should have a full nail. Reduced nail or absence of a nail is a defect.

The Silkie chicken should have a face with a smooth surface, with the skin being fine and soft in texture. The face, wattles and comb should be a dark mulberry in color, approaching black. The eyes should be large, round and prominent, solid black in color. The comb should be walnut shaped and be set firmly and evenly on the head, with the male's comb being larger than the female. A strawberry or cushion shaped comb is equally ideal by British Standards.

In bearded variety Silkies, the beard should be thin and full, extending back from the eyes and projecting from the sides of the face. The beard is composed of feathers turned horizontally backwards, from both sides of the beak and vertically downwards. The whole beard forms a collar of three ovals in a triangular group, giving a full ear muffing effect. A "Showgirl" Silkie is a newer variety which has a naked neck, but still has the crest on top of the head. Showgirl varieties are not recognized by American nor British Standards at this time.

The American Standard of Perfection is set by the American Poultry Association and the British Standard of Perfection is set by the Poultry Club of Great Britain. Table 4.1 provides details for both standards.

| Category | American Standard | British Standard |
| --- | --- | --- |
| Disqualifications | More or less than five toes. Shanks not feathered down outer sides. Feathers not truly silky (i.e. hard feathers), except in primaries, secondaries, leg, toe and main tail feathers. Vulture hocks (hard feathering in the hock which points down). Bright red comb, face or wattles. Any yellow skin. | More or less than five toes. Single comb. Featherless legs or feet. Vulture hocks (hard feathering in the hock which points down). Green, yellow, or white legs. |
| Recognized Colors | White, black, blue, partridge, buff, gray, splash, self-blue (lavender), and paint. | White, black, blue, gold, partridge. |
| Size | Cock: 36 oz. (2 lb. 4 oz.). Cockerel: 32 oz. (2 lb.). | Male standard: 64 oz. (4 lb.). Female standard: 48 oz. (3 lb.). |

|  | | |
|---|---|---|
|  | Hen 32 oz. (2 lb.). Pullet 28 oz. (1 lb. 12 oz.) | Male bantam: 22 oz. (1 lb. 6 oz.). Female bantam: 18 oz. (1 lb. 2 oz.). |
| Skin | Dark blue. Any yellow skin is a disqualification. | Dark, almost black. Green, yellow, or white legs are a disqualification. |
| Beak | Short and stout, curving to a point. Leaden blue in black and white varieties. Bluish black in blue and partridge varieties. Slaty blue in buff and grey varieties. | Short and neat. Slate grey (black in black variety only). Green in the beak is a defect. |
| Face | Surface smooth, skin fine and soft in texture, Deep mulberry color, approaching black in black and white varieties. A red face, comb, or wattle is a disqualification. | Color should be black in both sexes, tending to a deep mulberry in the male. A red face, comb, or wattle is a defect. |
| Eyes | Large, round, prominent. Black in color. | Black in color. |
| Wattles | Male non-bearded: medium size, concave, nearly round, fine in texture, free from wrinkles or folds. Male bearded: very small, concealed by beard, natural absence preferred. Female non-bearded: small, concave, forming a half circle, fine texture, free from wrinkles or folds. Female bearded: small to | Male: small and neat, dark mulberry in color. Female: nearly absent, black in color. A red face, comb, or wattle is a defect. |

| | | |
|---|---|---|
| | nonexistent, concealed by beard. Should be deep mulberry in color for both sexes, approaching black in black and white varieties. A red face, comb, or wattle is a disqualification. | |
| Comb | Male: walnut shaped, set firmly and evenly on head, almost circular in shape, preferably broader than longer, with a number of small prominences over it, a slight indentation of furrow, transversely across the middle, rising at a point just forward of the nostrils and extending backwards to a point parallel with the front of the eyes. Female: walnut shaped, very small, well formed. Rest of the description same as the male. Both sexes the comb should be a deep mulberry color, approaching black in black and white varieties. A red face, comb, or wattle is a disqualification. | Walnut, strawberry or cushion shaped. Comb should not be of the single type. Should be slightly rounded in the male (more wide than long) with an indentation running left to right and many small lumps across it (it has been likened to a flattened mulberry). In the female the comb should be very small and hidden by the crest (although the comb should not split the crest). Horns on the comb are a defect in either sex. A single type comb is a disqualification. A red face, comb, or wattle is a defect. |
| Crest | Male: medium size, soft and full, as upright as comb will permit, having a few silky feathers streaming gracefully backwards from | A pompom like a powder puff which in the male will have a shiny 'streamer' feathers pointing |

|  |  |  |
|---|---|---|
|  | lower and back part of crest. Female: medium size, soft and full, globular, upright, well balanced. | backwards. The crest should not obscure the eyes. No crest or a poor crest is a defect. Crest obscuring the eyes is a defect. Split crest is a defect. |
| Earlobes | Male non-bearded: small, oval, fine in texture, free from wrinkles or folds. Male bearded: very small, almost concealed by muffs. Female: very small, rest of description the same as the male. Turquoise blue in color. | Small and neat. Preference given to turquoise blue earlobes, although mulberry earlobes are permitted. |
| Head | Moderately small, short, carried so that a line drawn parallel with tip of the tail will bisect the comb. | Short and neat |
| Neck | Short, gracefully arched, with a very full hackle flowing well over the shoulders. | Short and neat |
| Back | Male: short, broad from shoulders to saddle, quite rounded its entire length rising gradually from middle of the back towards tail. Female: short, broad from shoulders to cushion, quite rounded its entire length, rising gradually from middle of the back towards tail. | Short, rising through to tail |

| | | |
|---|---|---|
| Tail | Male: short, very shredded at ends, well spread at base, filled underneath with an abundance of soft feathers which are overlapped by coverts and lesser sickles, the whole forming a duplex curve with back and saddle. Sickles, lesser sickles, and coverts- abundant, soft, well curved, without hard quills, concealing main tail feathers. Female: short, very shredded at ends, well spread at base, filled underneath with an abundance of soft feathers which are overlapped by cushion and coverts, the whole forming a duplex curve with back and cushion. | Short, rounded and fluffy. |
| Wings | Medium size, closely folded, carried well back and nearly horizontal, well above the lower thighs ending short of stern. Shoulders and fronts: concealed by hackles and breast feathers. Bows and coverts: very well rounded. Primaries: medium length, well shredded, tapering convexly to stern, tips concealed by saddle feathers. | Wings should be ragged with some of the flight feathers hanging down, almost looking tattered. Split wings (where there is a gap between the primaries and secondaries) are a defect. |

| | | |
|---|---|---|
| Breast | Carried forward, very full, well rounded and of great depth and width. | Broad and rounded |
| Body | Body of moderate length, broad, deep and well rounded from breast bone to stern and let down well between the legs. | Broad and stout. Overall shape is cobby and rounded. In the female the underfluff should nearly touch the ground. |
| Plumage | Silkie and fluffy with hair-like feathers throughout. The feathers should be profuse and thick, and cover the thighs, run down the shanks and cover the middle and outer toes. There should be an absence of hard feathering. Disqualifications: Shanks not feathered down outer sides. Feathers not truly silky (i.e. hard feathers), except in primaries, secondaries, leg, toe and main tail feathers. Vulture hocks (hard feathering in the hock which points down). | Silkie and fluffy with hair like feathers throughout. Entire body should be covered in abundant fluff. The feathers should be profuse and thick and cover the thighs, run down the shanks and over the middle and outer toes. There should be a lack of hard feathering. Hard feathers including tail feathers are a defect. Featherless legs or feet are a disqualification. Vulture hocks (hard feathering in the hock which points down) are a disqualification. |
| Beard and Muffs | Bearded varieties: thick, full, extending back of eyes and projecting from sides of face and composed of feathers turned horizontally | Bearded types should have full ear muffling and a beard which covers the lower part |

| | | |
|---|---|---|
| | backwards, from both sides of the beak, from the center, vertically downwards, the whole forming a collar of three ovals in a triangular group, giving a muffed effect. | of the face with reduced wattles. |
| Legs and toes | Male: legs short and stout, set well apart, straight when viewed from the front. Lower thighs short, stout at top, tapering to hocks, abundantly feathered. Hocks covered with soft and silky feathers curving inwards about the hocks. Shanks rather short, stout in bone, well feathered on the outer sides with silky plumage, the upper part growing out from thigh plumage and continuing into foot feathering. Spurs medium size and length, set just above the fifth toe. Five toes, the three front straight, well and evenly spread, the hind toe double, the normal toe in natural position and the extra toe placed above, starting from close to the other toe, but well formed, longer than the other toes and curving upwards and backwards; the outer and middle toe | Legs should be short and wide set but not bowed. They should be well fluffed at the thigh and moderately feathered along the shank (no hard feathers). The toes, which are also feathered, should be five in number, with the fifth toe coming off of the fourth. Pincer toes (like a crab claw) are a fault. Each toe should have a full nail. Legs and feet should be a dark slate color. Reduced toe size or absence of a nail is a defect. More or less than five toes is a disqualification in either sex. |

|  |  |  |
|---|---|---|
|  | well feathered. Female: same as male except no spur. A bare middle toe is a serious defect in either sex. More or less than five toes is a disqualification in either sex. |  |
| Saddle | Male: Rising from back at base of cape, very broad and round, plumage profuse and long, lower saddle feathers flowing over tips of wings and mingling with fluff. |  |
| Cushion | Rising from back at base of cape, very broad and round, plumage abundant. |  |

Table 4.1: American Standard of Perfection and British Standard of Perfection for Silkie chickens.

(American Poultry Association "Standard of Perfection" 2015 edition. For full standard you may purchase this publication at http://wwwamerpoultryassn.com)

(Poultry Club of Great Britain "British Poultry Standards" 2017 edition. For full standard you may purchase this publication at http://www.poultryclub.org)

Obviously, it would be difficult to find a perfect specimen. As we will discuss in the breeding chapter, a key to breeding Silkies as show birds is to make up for any defects found in one sex in the other sex. This will improve the quality of your show bird flock over time.

You do not have to meet every Standard of Perfection to enter a Silkie chicken in an exhibition. You can still win prizes if it is one of the best looking Silkies there. You will be far ahead of many people by understanding and focusing on the standards. If you want to show your Silkies, just do your best to breed them for these qualities and feed them a healthy diet. Your diligence will pay off because you will be competing with many people who randomly buy chicks or just use a hit and miss method of breeding. The quality of your Silkies will be much better than theirs if you concentrate on these standards.

# Chapter 5: Breeding Silkie Chickens

Breeding Silkie chickens is not difficult at all, because the hens make excellent mothers. They will set on their eggs and often go broody. This makes egg collection difficult at time but makes breeding a breeze! In fact, Silkie hens are frequently used as sitters to hatch the eggs of other breeds because of this brooding characteristic. This can be a helpful trait if you have a mixed breed flock or multiple flocks. Many people like to cross breed Silkies with other breeds to mix in the brooding quality of Silkies while maintaining some of the benefits of the original breed.

Not only are Silkies excellent setters, seldom breaking or soiling an egg, but they make the most wonderful mothers, affording their chicks greater protection from prowling cats and dogs than any bantam hen. I've actually seen a Silkie hen with a brood of chicks drive a stray dog out of the yard in the most business-like manner imaginable! They do not range far, and their soft, fluffy plumage provides ideal shelter for chicks. If I were a baby chick, it would be hard to imagine anything nicer than a Silkie mother to snuggle up to.

While Silkie hens are unsurpassed as mothers, there is a natural genetic failing that occurs from time to time. The excretions from the baby chick attaches to the silk feathers of the mother, forming a noose which surrounds the baby chick's neck, and they are hanged. There is not much you can do about this if you want to let your Silkie hens raise their own chicks, it just happens from time to time.

During breeding season, you can take the selected rooster away from the hens in the evening and feed him all the corn he can eat. (You can buy large bags of cracked corn almost anywhere that sells chicken feed). Then put him back with the hens in the morning. This routine will help guarantee fertilized eggs, although it isn't necessary if you don't have space or don't feel up to it. Always make sure there is a clean, comfortable nest, secluded and not exposed to noises and vibration to encourage successful breeding.

If you are breeding your Silkie chickens just to keep as pets and for egg production, you will not be quite as concerned with breeding for specific characteristics. In this case a Silkie rooster can mate with 5-6 hens. If you are breeding Silkies as show birds, you will want to pick 2-3 of your best hens to mate with your best rooster. Be sure to pay attention to the Standards of Perfection for show birds as described in the previous chapter.

Never breed a Silkie with a disqualification as a show bird. See the list of disqualifications in Table 4.1. Feathering, color, size and shape are all important qualities to consider when selecting show birds. Primary ideals to breed Silkies for are silky plumage that should cover the shanks and nearly touch the ground, broad shape across the back, full breast, a dark (darker the better) face, comb, and wattles, blue ear lobes, and five toes. A well-formed pompom like crest is also important, and the shape of combs is also an important consideration.

A walnut comb is preferred by American standards while a walnut, strawberry or cushion comb is preferred by British standards. Never select a Silkie with a single type comb to breed as a show bird. A single type comb is a disqualification by British standards and a serious defect by American standards. Whenever possible, avoid a poor comb in either sex. However, do not avoid breeding an otherwise desirable show bird just because of a poor comb (unless that comb is of the single type). Feathering, crest, color, size and shape are all more important. If you have a couple birds to choose between with equal of these most important qualities, choose the one with the good comb.

Do not breed birds with color defects for show. Skin color should be dark blue to black. Never select a Silkie with yellow or green skin for breeding. See the previous chapter for specifications on different color varieties and pay attention to recommended standards for feathering color. For example, a white Silkie should ideally have white feathering all over without any traces of other colors.

Make sure that shanks and feet are well feathered for breeding stock and do not breed Silkies that have hard feathers. Some Silkies are born without feet feathers and ideally should not be selected for breeding. Particular attention when mating should be paid to the male's tail coverts and wing bows, as these two places are hard to get silky. Common shape faults include the body and back being too long, being too high on the legs (i.e. too leggy), lack of front (i.e. breasts not full), and pinched tails.

The male Silkie is more likely to influence the color and shape while the female Silkie is more likely to influence the size and the feather texture. The key to breeding Silkies as show birds is to pay close attention to the Standards of Perfection and to make up for any defects found in one sex in the other sex. For example, if you have a rooster with a poor comb that is a bit too skinny, but he has excellent color, feathering and shape, choose a large hen with a good comb.

In bearded varieties, if you are trying to increase the size of the beard, work on reducing the size of the wattles and the large beard will fill in naturally. Similarly, if you want to increase the size of the Silkie's crest, work on reducing the size of the comb and the crest will fill in naturally.

Do not take a good rooster and mate him indiscriminately with a number of hens. This hit and miss method will usually produce a lot of misses! If you choose your two or three best hens to mate with him, you will be pleased with the chicks that result. If you have a large flock, you can mate up to five or six hens with one rooster. Just make sure you are choosing high quality hens. Luckily, Silkies breed very true to both type and color, making an easy guess of what type of chick you will get based on the parents.

We discussed how Silkie hens make great mothers, but even so some people prefer the control of incubating eggs themselves and raising the chicks indoors until they are large enough to go outside. Although it is lovely to see Silkie hens raising their chicks in mother nature's way, there is a greater rate of fatality with this method. If you incubate the eggs yourself, you can sometimes have a 100% success rate with chick survival. The incubation period will be about 21 days for Silkie eggs.

When starting your flock, you can buy Silkie hatching eggs and an incubator instead of buying chicks. Or you can buy chicks to initially start your flock and incubate the eggs they lay later. This is of course more work and the incubator an investment, but it is fun and rewarding to hatch chicks this way. It is also a great hands-on learning activity for children.

If you are going to incubate eggs yourself, you may want to check out this guide to building a high-quality egg incubator that is guaranteed to yield a high hatching rate. You can build this incubator using cheap parts from your local hardware store even if you have no experience in building things. You will not have to worry about turning the eggs, the incubator you build will do this automatically. Visit http://www.incubatormaker.com/?hop=chicken121 for details (this is an affiliate link). Your incubator will be just as good as the high-end versions used in professional hatcheries, just on a smaller scale. And building the incubator will make the project even more educational if you are doing it with your children.

If you choose to purchase an incubator, check your local agricultural supply store. Most also have websites if there is not a store near you. Of course, you can find them on Amazon as well. You can spend anywhere from $50 - $800 on an incubator depending on the level of automation (whether you turn the eggs yourself) and how many eggs it holds. Commercial sized units can go much higher in price, but this is an average range of pricing for a personal homestead sized incubator. If you go toward the lower price range, plan to turn the eggs three times per day.

Whether you buy Silkie chicks or hatching eggs, read reviews from other customers and *buy the best*. The cost of feeding is the same and the only difference is the initial purchase price. Especially with

Silkies where the standards are fairly specific, you want to have good birds.

You will want to introduce new Silkie chickens into your flock about every three generations. This will prevent problems associated with inbreeding. If you are concerned with consistency of characteristics, you should go back to the same hatchery you originally bought chicks or hatching eggs from if possible.

It is difficult to distinguish sex in young Silkie chicks as compared to other breeds. There are a couple of telltale signs. The comb of the male is much larger than the female. The wattles of the female are very small and oval shape, and sometimes she is practically devoid of them. If you can't make out the difference, you'll just have to wait for time to tell!

If you've incubated and hatched eggs, you will want to keep young Silkie chicks in a cage with the bottom lined with bedding after they are past the heat lamp stage. If the weather is nice the cage can be kept outside. Silkie chicks can safely be introduced into the pen with the other chickens by 12-14 weeks of age. You can put them outside during the day a couple weeks earlier if they have a separate fenced area next to the main pen. This will help the chickens get used to each other without endangering the chicks. Don't let young Silkie chicks get wet. Wet beaks literally cause them to catch a cold, which could spread through your flock or develop into something more serious.

Whether you allow the hens to hatch their eggs or incubate them yourself, watching Silkie chicks grow up is certainly a pleasure!

## Chapter 6: Silkie Healthcare

Unfortunately, Silkies are prone to certain poultry diseases more often than other breeds. Potentially fatal diseases and conditions that Silkies are prone to include Marek's Disease, Coccidiosis, and water on the brain. Other conditions that Silkies are prone to include feather mites and Scaly Leg. It can be said that Silkies are a bit higher maintenance than other breeds in the area of healthcare. However, with proper preventative care most Silkies can live long healthy lives!

It is highly recommended to vaccinate Silkie chicks against Marek's Disease. Any poultry can contract Marek's Disease, but Silkie chickens are highly susceptible to the disease as compared to other breeds of chicken. Without going into too much detail, the disease is caused by a type of herpes virus (a poultry virus that does not affect humans). Some birds will not become ill and will show no signs of the disease but will be carriers. Most birds contracting Marek's Disease will develop a type of cancer due to the virus. For approximately seventy percent of birds with Marek's Disease, it is ultimately fatal. Symptoms of the disease depend on the location where tumors grow and can range from head spasms to leg paralysis. The disease typically appears between six and thirty weeks of age.

Luckily, there are vaccines available for Marek's Disease. The vaccines can be given within one day of hatching or when the chick is fully formed but still in the egg. You can administer the vaccine within three days of hatching, however within one day is recommended.

While the vaccination does not prevent the chicken from contracting Marek's Disease and becoming a carrier, it does prevent them from developing illness and cancer due to presence of the virus. The effectiveness rate of the vaccine has been shown to be over ninety percent.

If you are purchasing chicks, select a hatchery that offers Marek's Disease vaccination. It is probably change per chick to add this important vaccination. If you are hatching your own eggs this becomes more complicated. You can purchase a certain type of Marek's disease vaccine (sero type 3) with no license or prescription, but unfortunately you will probably have to purchase more quantity of the vaccine than you want because it is only sold in large quantities. However, the vaccine will keep in the freezer, so you can have enough for many hatching sessions to come. Once thawed the vaccine must be used within one hour. Try not to be squeamish about giving the chicks the vaccine. It is not difficult to administer and will become second nature after the first couple rounds of chicks. Watch a couple videos to see how it's done or let a friend with experience help you out the first time. Of course, you could just risk it and skip the vaccine, it is a personal decision. When people first started with Silkie chickens, the vaccine did not exist!

Coccidiosis, which is a condition of the intestines, affects Silkie chicks in particular. Coccidiosis is an intestinal disease that occurs when a microscopic parasitic organism (called a protozoa) attached itself to the intestinal lining of a chicken, prevents the chicken from absorbing nutrients, and creates an environment in which bacteria can thrive. It is transmitted through the droppings of other birds, so great attention must be paid to cleanliness and hygiene. Coccidiosis is usually fatal if not treated.

The potential for Coccidiosis contraction is one of the reasons many breeders rear Silkie chicks inside in box cages and change the litter quite frequently. Once fully feathered the chicks can be put outside, but on clean ground and preferably not in wet conditions because this seems to aggravate the condition. Of course, you can let Silkie hens raise their own chicks, but take extra precautions to keep the coop area clean and dry if you do.

If you are purchasing your chicks through a hatchery, some offer a vaccine for Coccidiosis. Coccidiosis vaccine is like a flu vaccine for people, it is not one hundred percent effective. Unlike vaccine for Marek's Disease which almost all hatcheries offer, some hatcheries do not offer a vaccination for Coccidiosis. If the chicks you purchase are not vaccinated or if you are hatching your own chicks, give medicated chick starter feed for the first eight weeks which contains Coccidiostat, and this will be nearly as effective as the vaccine (there is not a similar option for prevention of Marek's Disease). Do not give vaccinated chicks the medicated feed; this will simply cancel out the vaccine.

It is possible for an adult chicken to contract Coccidiosis, although they are not nearly as susceptible as chicks. To prevent Coccidiosis, make sure water is clean and fresh, keep feeding areas clean and dry, and ensure that your chickens are not overcrowded. You may also want to quarantine any new additions to your flock for two weeks, to prevent the spread of not only Coccidiosis but other common poultry diseases.

It is very difficult to confirm a case of Coccidiosis without a veterinarian's diagnosis, because the disease has symptoms which appear similar to other conditions. A veterinarian can test a stool sample to determine if a chicken is infected. Symptoms can present themselves either gradually or suddenly and may include blood or mucous in the droppings, weak chickens not moving around much, huddling together as if cold, pale comb and skin, loss of appetite, ruffled feathers, weight loss, baby chicks failing to grow, inconsistent egg laying or not at all, and diarrhea. Again, the tricky part is that these can all be signs of other diseases.

If Coccidiosis is confirmed, the infected birds should be quarantined if it has not spread to the whole flock. Clean out the coop and run area thoroughly. There is a commercial treatment available which blocks the ability of the parasite to uptake and multiply. The medication is called Amprolium and is available over the counter. It is a liquid that can be added to the water. However, if the chicken is already very sick you may need to administer the medication orally, because they may not be drinking much water. The medication should be administered daily for seven days. For serious infections the cycle may need to be repeated for a total of two weeks treatment time. Just remember, a dry, clean living area with drinking water being changed daily is the best preventative for Coccidiosis.

Many Silkie chicks are born with a dome on their skull which will later produce an excellent crest. It appears as a large bump on top of the chick's head. Water on the brain is a particular problem with crested breeds of chickens (although not limited to these breeds). Water on the brain is often mistaken for Marek's Disease. An infection in the enlarged cranial cavity produces fluid which in turn presses on the brain. The main symptom of water on the brain is walking backwards and falling over. The chick may spin around in circles and then suddenly recover.

Treatment for water on the brain is a bit time consuming but usually very effective. Unfortunately, the antibiotic and anti-inflammatory drug mixture needs to be prescribed by a veterinarian; it is not available over the counter. The mixture is injected into the breast muscle of the chick daily for four weeks. Your vet can show you how to administer the injection. In addition, the chick must be isolated and administered a liquid feed (chick pellets ground up with water) via a large syringe directly into the mouth at regular two to four-hour intervals during the day, with the last feed at night. Again, while time consuming this treatment for water on the brain is highly effective.

Because of their fluffy feathers, Silkies suffer from feather mites more often than other chickens. Cleanliness in the pen is the best preventative for feather mites. You can also treat the coop and run area with a good poultry mite powder. However, it is like having fleas in cats and dogs. Sometimes it just happens if the mites are present in your area. You can sprinkle the poultry mite powder directly on your Silkies if mites are found. The powder is a good interim solution between baths. The best cure for feather mites is weekly baths in a flea and tick shampoo. Shampoos marketed for dogs are safe for your Silkies and the same elements which kill fleas and ticks will kill feather mites.

As with fleas, feather mites are more easily eradicated if they are found before there is a total infestation. Check your Silkies for small black mites, which can be difficult to see because of their dark skin. Also check their nesting boxes. You can sometimes see the little mites jumping around if present. Clean out all nesting material and replace with fresh if mites are found. After the bath in medicated shampoo, you can apply a small amount of Sulphur ointment to the Silkie's skin to help repel mites. Sulphur ointment is available over the counter at most pharmacies, as it is also used in a variety of human applications.

A more serious skin problem in Silkies is Scaly Leg, caused by a small parasite that gets under the scales of the leg and disfigures the bird. The scales become raised and uneven and leg feathering suffers. Scaly leg is incredibly itchy and uncomfortable to the chicken. Symptoms include raised leg scales, white salt crusts around the legs, discomfort in walking, bleeding on the legs, legs appearing swollen, and impaired circulation causing toe loss or foot deformity.

Silkies are susceptible to this condition because the feathering on their legs encourages the accumulation of dirt. The best way to avoid the problem is strict cleanliness of legs and feet, washing regularly (say once or twice per month) with warm water containing liquid soap and a mild disinfectant. Other preventative measures include treating the coop and run area with a good poultry mite powder, and oiling perches with vegetable oil.

Silkies contracting Scaly Leg are generally unfit as show birds and become carriers of the condition. To treat Scaly Leg, for the first week or so, soften the crusts with petroleum jelly, coconut oil, or baby oil applied to the scales (not too much or it will get messy). This will help sooth the skin and suffocate the mites. Then give the legs and feet a good soak in hot water containing flea and tick shampoo or liquid soap mixed with a mild disinfectant. Use a soft toothbrush to loosen the deposits but stop if blood starts to show. Finally, dunk each leg in Surgical Spirit for thirty seconds. Surgical Spirit is mostly composed of alcohol and can be purchased over the counter.

Optionally, you can then add a thin coat of Sulphur ointment mixed with medicated tar to the legs. Medicated tar is available over the counter and has uses for human skin conditions as well, but use is controversial because the tar has been shown to contain carcinogens. Sulphur ointment mixed with medicated tar has been a classic treatment for Scaly leg in poultry dating back to the 1800s, prior to the discovery of carcinogens. Whether to perform this step in the treatment regiment is a personal decision. Either way, repeat the treatment every week until symptoms of Scaly Leg clear up. It may take awhile for the legs to look better, often not until the next annual moult, but be assured they will eventually improve.

# Chapter 7: Silkie Chickens for Eggs and Meat

The primary purpose of Silkie chickens is generally as pets and show birds, but most often a secondary purpose is egg production. You can find breeds better suited for egg production if that is your primary goal, or a utility breed that is good for both egg and meat production. However, if you are wanting to keep Silkies anyway, you might as well enjoy their eggs!

You can count on Silkie hens to lay between 80 – 120 small (just over two ounces) eggs per year. (Comparatively, many utility breeds lay 200 – 250 eggs per year while breeds specialized in egg production lay 250 – 350 eggs per year). Most Silkie hens will lay 3 – 5 eggs per week, except for the long break they take in the fall for molting (5-8 weeks average time). You could also get a shorter soft molting break in springtime.

The color of the eggs varies from porcelain white to light brown to pinkish. As with other small and bantam breeds, the eggs have a higher yolk to white content. Which is great with me because the yolk is my favorite part! Yolks are generally dark yellow, almost orange, and very rich. Of course, they taste much better than store bought eggs.

Laying is often interrupted by the tendency to go broody in Silkie hens. This can also make egg collection a pain at times. Be sure to collect the eggs promptly at the same time everyday unless you are trying to allow the hen to set on the nest and hatch chicks. This will help interrupt the tendency for broodiness. If you notice a problem, you could even check for eggs twice per day. While the broodiness tendency is a pain when it comes to collecting eggs, it is certainly helpful when it comes to breeding.

Silkie hens start laying between 7 – 12 months of age, with 9 months being about average. This is much longer than most breeds take so be patient. The longer it takes a hen to start laying, the more eggs she will produce for you when she does start. If a hen over one year old is still not laying eggs, it could be a calcium deficiency. Try providing some crushed oyster shell. Also make sure that the hen has a healthy diet. Remember that Silkies must be supplemented by fresh produce and protein and cannot be healthy on commercial chicken feed alone. Silkie hens will lay steadily for 3 - 4 years, then occasionally until not at all.

Silkie chickens will continue to lay over winter, unlike many small or bantam breeds. As discussed previously, Silkies cannot tolerate climates with extreme winters, but they can certainly tolerate temperate climates that have some winter days which drop below freezing with snow fall. If you keep your Silkies dry, including underfoot, they will stay happy and you can collect eggs even in the snow. Cod liver oil is a good supplement to encourage winter laying.

While considered a delicacy in many Asian cultures, Silkies are not generally used for meat in western cultures because of the black skin and bones, not to mention the black blood veins. The other thing is that Silkies become cherished family pets so many people just couldn't bring themselves to kill them! Some people just find them to be too cute to eat.

However, if you are open to using your Silkies for meat, they make delicious table birds. The meat is of the whitest and finest in texture. It is more flavorful than most chicken. While a prepared Silkie does not make so attractive an appearance as some other breeds, if tried it is usually relished. In my opinion it is worth giving a try at least once, but I love to try new things!

Like using Silkies for egg production, it is not worth it for your primary purpose of keeping Silkies to be meat production. They just take too long to grow out and never get very big even at full grown. However, it is a nice secondary purpose to harvest Silkies for meat and is also a unique dining experience. The ideal time to slaughter a Silkie chicken is between 6 – 12 months of age, but up to three years of age is fine. Older than that and the meat will be tough, but you can get away with slaughtering an older chicken and boil the meat to increase tenderness.

The breasts are fairly small in Silkies, as is the case with all breeds that are not selected for meat production. You can expect to get 1 – 1.5 pounds of meat per Silkie chicken. A three-pound bird will yield about 1.5 pounds of meat. With a larger Silkie cock you may get up to two pounds of meat. It is a bit more of a pain to pluck Silkie feathers than with other breeds (because of the lack of barbicels) and some people prefer to dry pluck Silkies for this reason.

The light to dark meat ratio in Silkies is the same as regular chickens. Cook Silkies just like you would any other chicken! It's just like any other chicken other than pigment and as bony as any other bantam you may eat. You can eat the skin of the Silkie, but normally it is removed before serving. In some Asian cultures and especially in China, the Silkie is boiled to make chicken soup for medicinal purposes. Like eating chicken noodle soup when you are feeling under the weather, but Silkies are said to have more health benefits and contribute to faster healing. Lots of fresh herbs are stuffed in the cavity when the bird is boiled to make this medicinal soup.

Speaking of medicinal values, an interesting thing about Silkie meat is the high levels of carnosine, a naturally occurring peptide which is sold as a dietary supplement. People take it to increase muscle mass, ward off the effects of aging, and to alleviate symptoms of diseases such as diabetes or autism. Studies have shown that the black skinned chicken is one of the richest natural sources of carnosine.

Some people like the Silkie's taste and the black skinned chicken health benefits so much that they have developed a strategy to cross Silkies with broiler breeds to increase the harvest size while maintaining the black skin and bones. These crosses are referred to as Silkie broilers, Taihe chickens, or black meat chickens (although the term black meat chickens extends to multiple breeds). Taihe chickens are a popular meat chicken in many Asian markets.

# Chapter 8: Conclusion – Are Silkies the Breed for You?

Silkies are one of those special breeds of chickens. If you bought this book you probably have already decided, you want them! They are among the friendliest breeds of chickens you can get. As mentioned previously Silkies are cuddly and enjoy sitting with you. Many breeds tolerate the attention, but Silkies really seem to relish it and want to be pets. They are one of the best breeds you can have around children. Silkies are beautiful to look at in the backyard and so fun to enter in shows and exhibitions.

However, if you fall into one of the following categories, Silkies might not be for you. Perhaps you are mainly concerned with egg production and making an income selling those eggs. You would want a breed selected for laying the maximum number of eggs, such as the Leghorn. If your prime objective is raising chickens for meat, you might prefer a faster growing breed for broiler production such as a Cornish breed, or a good utility breed (good for egg production and meat), which has the more typical white or yellow skin color.

Typically, people have already decided at this point that they want to keep Silkies as pets and/or show birds. However, if you live in a northern climate with extreme frigid winters, Silkies are probably not for you unless you want to invest in a safe heated habitat, as discussed in the housing chapter. Silkie chickens can tolerate cold better than many small or bantam breeds and it is not a problem to keep them in temperate climates that experience some below freezing temperatures and snow fall. However, if you regularly have temperatures zero degrees and below, your Silkies will almost certainly die unless kept warm. There are many other breeds of chicken which can tolerate these extreme temperatures with no need for artificial heating.

If you live in a tropical climate or any climate type with very high levels of rainfall, you will need to consider how to keep your Silkies dry, including underfoot, should you choose to keep them. There are many other breeds that can tolerate wet conditions; Silkies are not one of them. Silkies tolerate heat very well, making them an ideal choice in both hot, dry climates and in temperate climates.

As discussed in the healthcare chapter, there are a few diseases and conditions that Silkies are more prone to than other chicken breeds. This doesn't present a problem for most owners but recognizing that Silkies do require a bit more preventative care than other breeds is important.

If the desire for specialization in egg or meat production, the bit of extra attention required, and the climatic issue do not present a problem, the Silkie is going to be a great breed for you. They are the perfect pet and show bird breed. You can count on them to be friendly, and they are beautiful birds to observe. Once you start with Silkies, they will be a treasured part of your flock for years to come!

Whether or not you decide to raise Silkies, do consider a heritage breed of chicken anytime you are adding to your flock. Many heritage chicken breeds are still endangered due to preference for a few fast-growing breeds that suit the needs of factory farms. According to the Livestock Conservancy, a heritage chicken is one that is recognized as an American Poultry Association Standard Breed and has been since prior to mid-twentieth century, is naturally mating, has a long productive outdoor lifecycle, and has a slow growth rate. These slower growing heritage chickens have better tasting eggs and meat. And by owning them you get to help an endangered species and diversify the genetic pool of chicken breeds. For more information, visit the Livestock Conservancy's website at: https://livestockconservancy.org/index.php/heritage/internal/heritage-chicken.

Made in the USA
Monee, IL
29 July 2024

62937209R00042